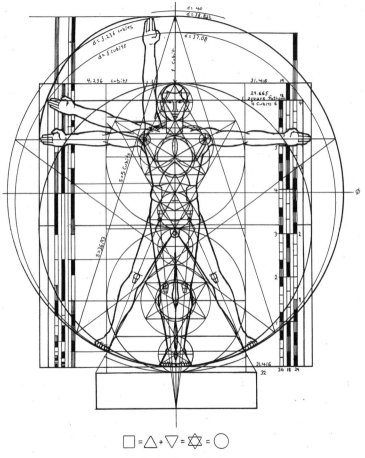

Vitruvian Man (Harding)

Published by
Walker Publishing Company, Inc., New York
Distributed to the trade by
Holtzbrinck Publishers

Printed on recycled paper.

Library of Congress Cataloging-in-Publication Data
has been applied for.

ISBN-10: 0-8027-1539-7
ISBN-13: 978-0-8027-1539-5

Visit Walker & Company's Web site
at www.walkerbooks.com

First U.S. edition 2006

1 3 5 7 9 10 8 6 4 2

Designed and typeset by
Wooden Books Ltd, Glastonbury, UK

Printed in the United States of America

THE GOLDEN
SECTION
NATURE'S GREATEST SECRET

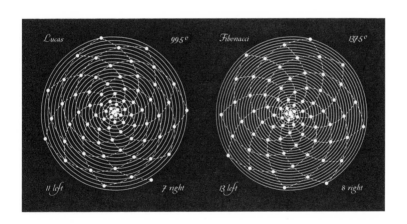

Scott Olsen, Ph.D.

Walker & Company
New York

Deep gratitude to my loving parents, Ilene & Clarion

Thanks to contributors: Keith Critchlow, John Michell, Lance Harding, Benjamin Bryton, Garth Norman, Mark Reynolds, Robin Heath, Richard Heath, Pablo Amaringo, Zachariah Gregory, & especially to my editor John Martineau. I am grateful for discussions with: Dan Pedoe, David Bohm, Huston Smith, Douglas Baker, Stephen Phillips, Edgar Mitchell, David Fideler, Garileo Pedroza, Robert Powell Sr., Alexey Stakhov, Michael Baron, & Bill Foss. Special thanks to my lovely wife, Pam. Thank you CFCC for the sabbatical. Further sources: P. Hemenway, Divine Proportion; G. Doczi, Power of Limits; M.Schneider, Golden Section Workbook; Kairos-foundation Φ worksheets; M. Livio, Golden Ratio; M. Ghyka, Geometry of Art & Life; H.E. Huntley, Divine Proportion; R.A. Dunlap, The Golden Ratio.

Above: Francino Gaffurio's early woodcut of a Liberal Arts lesson. Previous Page: Lucas and Fibonacci spiral phyllotaxis showing Archimedean spiral (after Bursill, Rouse & Needham).

CONTENTS

INTRODUCTION

Nature holds a great mystery, zealously guarded by her custodians from those who would profane or abuse the wisdom. Periodically portions of this tradition are quietly revealed to those of humanity who have attuned their eyes to see and ears to hear. The primary requirements are openness, sensitivity, enthusiasm, and an earnestness to understand the deeper meaning of nature's marvels exhibited to us daily. Many of us tend to walk through life half asleep, at times numbed, if not actually deadened to the exquisite order that surrounds us. But a trail of clues has been preserved.

The secretive tradition centers on a study of number, harmony, geometry and cosmology that stretches back through the mists of time into the Egyptian, Babylonian, Indian and Chinese cultures. It is evident in the layout and relationships of the stone circles and underground chambers of ancient Europe, as well as in Neolithic stones discovered in Britain, fashioned in the form of the five regular solids. There are further clues in Mayan and other Mesoamerican artifacts and buildings, and across the ocean the Gothic masons embedded them in their cathedral designs.

The great Pythagorean philosopher, Plato, in his writings and oral teachings, hinted, though enigmatically, that there was a golden key unifying these mysteries.

Here is my promise to you: if you are willing to proceed step by step through this compact little book, it will be well nigh impossible not to grasp by the end a satisfying and stunning glimpse, if not deeply provocative insight, into Nature's Greatest Secret.

THE MYSTERY OF PHI
the golden thread of perennial wisdom

The history of the golden section is difficult to unravel. Despite its use in ancient Egypt and the Pythagorean tradition, the first definition we have comes from Euclid [325-265 BCE], who defines it as the division of a line in extreme and mean ratio. The earliest known treatise on the subject is *Divina Proportione* by Luca Pacioli [1445-1517], the monk drunk on beauty, and illustrated by Leonardo Da Vinci, who according to tradition coined the term *sectio aurea*, or "golden section." However, the first published use of the phrase occurs in Martin Ohm's 1835 *Pure Elementary Mathematics*.

There are many names for this mysterious section. It is variously called a golden or divine ratio, mean, proportion, number, section or cut. In mathematical notation it goes by the symbol τ, "*tau*," meaning "the cut," or more commonly Φ or ϕ, "*phi*," the first letter of the name of the Greek sculptor Phidias, who used it in the Parthenon.

So what is this enigmatic cut, and why is there so much fascination about it? One of the eternal questions asked by philosophers concerns how the One becomes Many. What is the nature of separation, or division? Is there a way in which parts can retain a meaningful relationship to the whole?

Posing this question in allegorical terms, Plato [427-347 BCE] in *The Republic* asks the reader to "take a line and divide it unevenly." Under a Pythagorean oath of silence not to reveal the secrets of the mysteries, Plato posed questions in hopes of provoking an insightful response. So why does he use a line, rather than numbers? And why does he ask us to divide it unevenly?

To answer Plato, we first must understand ratio and proportion.

RATIO, MEANS & PROPORTION
continuous geometric proportion

Ratio (*logos*) is the relation of one number to another, for instance 4:8 ("4 is to 8"). However, proportion (*analogia*) is a repeating ratio that typically involves four terms, so 4:8 :: 5:10 ("4 is to 8 is as 5 is to 10"). The Pythagoreans called this a four-termed discontinuous proportion. The invariant ratio here is 1:2, repeated in both 4:8 and 5:10. An inverted ratio reverses the terms, so 8:4 is the inverse of 4:8, the invariant ratio now 2:1.

Standing between the two-termed ratio and the four-termed proportion is the three-termed mean in which the middle term is in the same ratio to the first as the last is to it. The geometric mean between two numbers is equal to the square root of their product. Thus, the geometric mean of, say, 1 and 9 is $\sqrt{(1 \times 9)} = 3$. This geometric mean relationship is written as 1:3:9, or, inverted, as 9:3:1. It can also be written more fully as a continuous geometric proportion where these two ratios repeat the same invariant ratio of 1:3. Thus, 1:3 :: 3:9. The 3 is the geometric mean held in common by both ratios, binding, or interlacing them together in what the Pythagoreans called a three-termed continuous geometric proportion.

Plato holds continuous geometric proportion to be the most profound cosmic bond. In his *Timaeus* the world soul binds together, into one harmonic resonance, the intelligible world of forms (including pure mathematics) above, and the visible world of material objects below, through the 1, 2, 4, 8 and 1, 3, 9, 27 series. This results in the extended continuous geometric proportions, 1:2 :: 2:4 :: 4:8, and 1:3 :: 3:9 :: 9:27 (*see opposite*).

4

Ratio: between two numbers *a* and *b*

Ratio between *a* and *b* $a : b$ or a/b

Inverse ratio $b : a$ or b/a

Means: *b*, between *a* and *c*

Arithmetic Mean *b* of *a* and *c* $b = \dfrac{a + c}{2}$

Harmonic Mean *b* of *a* and *c* $b = \dfrac{2ac}{a + c}$

Geometric Mean *b* of *a* and *c* $b = \sqrt{ac}$

Proportion: between two ratios

Discontinuous (4 termed)
$a : b :: c : d$
  e.g., $4 : 8 :: 5 : 10$
  has invariant ratio $1 : 2$

Continuous (3 termed)
$a : b :: b : c \Rightarrow a : b : c$
  note *b* is the geometric
  mean of *a* and *c*

Plato's World Soul:

Extended continuous geometric proportion

$1 : 2 :: 2 : 4 :: 4 : 8$
  invar. ratio $1 : 2$
    or $1/2$

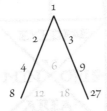

$1 : 3 :: 3 : 9 :: 9 : 27$
  invar. ratio $1 : 3$
    or $1/3$

Lambda diagram

PLATO'S DIVIDED LINE
knowing precisely where to cut

So, returning to our puzzle, why does Plato ask us to make an uneven cut? An even cut would result in a whole : segment ratio of 2:1, and the ratio of the two equal segments would be 1:1. These ratios are not equal and so no proportion is present!

There is only one way to form a proportion from a simple ratio, and that is through the golden section. Plato wants you to discover a special ratio such that *the whole to the longer equals the longer to the shorter.* He knows this would result in his favorite bond of nature, a continuous geometric proportion. The inverse also applies, *the shorter to the longer equals the longer to the whole.*

And why a line, rather than simply numbers? Plato realized the answer is an irrational number that can be geometrically derived in a line, but cannot be expressed as a simple fraction *(see page 54)*.

Solving this problem mathematically, and assuming the mean (longer segment) is 1, we find the greater golden value of 1.6180339... (for the whole), and the lesser golden value 0.6180339... (for the shorter). We term these Φ "fye" the Greater and ϕ "fee" the Lesser respectively. Notice that both their product and their difference is Unity. Furthermore, the square of the Greater is 2.6180339, or $\Phi+1$. Notice also that each is the other's reciprocal, so that ϕ is $1/\Phi$.

In this book we will generally speak of the Greater as Φ, the mean as Unity (1), and the Lesser as $1/\Phi$.

Notice *(below left to right)* that Unity can act as the Greater (whole), Mean (longer segment) or as the Lesser (short segment).

$1/\phi$	$1/\phi^2$		1	$1/\phi$		ϕ	1
1			ϕ			ϕ^2	

PHI ON THE PLANE
pentagrams and golden rectangles

Moving from the one-dimensional line onto the two-dimensional plane, the golden section is not difficult to discover.

Starting with a square, an arc centered on the midpoint of its base swung down from an upper corner easily produces a large golden rectangle (*below left*). Importantly, the small rectangle which we have added to the square is *also* a golden rectangle. Continuing this technique creates a pair of these smaller golden rectangles (*opposite top left*). Conversely, removing a square from a golden rectangle leaves a smaller golden rectangle, and this process can be continued indefinitely to produce a golden spiral (*opposite lower right, & cover*).

The golden section, which as we have seen unifies parts and whole like no other proportion, is intimately involved with the natural geometry of the pentagram (*opposite lower left*), the very emblem of life. Every point of intersection creates lengths which are in golden relationships to one another. An arm of a pentagram contains the key to another golden section spiral as a continuous series of increasing or shrinking golden triangles (*opposite top right*).

The golden cut of a line may be achieved by building a double square on the line and following the diagram (*below right*).

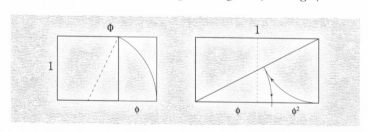

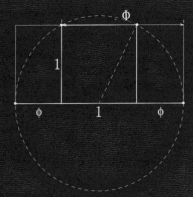

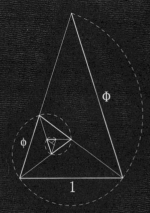

the Lesser and Greater derived from a square

the golden triangle

The basic operation of the golden section on the plane, showing features of golden rectangles, golden triangles, and the **Φ** : 1 relation between the diagonal of a pentagram and the edge of its enclosing pentagon. See if you can work out what the two unlabelled measures are in the diagram below.

removing squares

using rabatment to create a grid

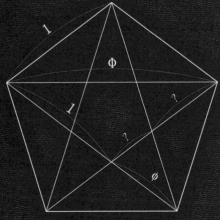

the golden section in the pentagram

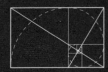

finding the occult center

THE FIBONACCI SEQUENCE
stepping stones to gold

Nature widely expresses the golden section through a very simple series of whole numbers. The astounding Fibonacci number series: 0, 1, 1, 2, 3, 5, 8, 13, 21, 34, 55, 89, 144, 233, 377... is both additive, as each number is the sum of the previous two, and multiplicative, as each number approximates the previous number multiplied by the golden section. The ratio becomes more accurate as the numbers increase. Inversely, any number divided by its smaller neighbor approximates Φ, alternating as more or less than Φ, forever closing in on the divine limit (*opposite lower right*). Each Fibonacci number is the approximate geometric mean of its two adjacent numbers (*see Cassini formula, page 53*).

Although officially recognized later, the series appears to have been known to the ancient Egyptians and their Greek students. Ultimately Edouard Lucas in the 19th century named the series after Leonardo of Pisa [c. 1170-1250], also known as Fibonacci (son of the bull), who made the series famous through his solution of a problem regarding the breeding of rabbits over a year's time (*right*).

Fibonacci numbers occur in the family trees of bees, stock market patterns, hurricane clouds, self-organizing DNA nucleotides, and in chemistry as with the uranium oxide compounds U_2O_5, U_3O_8, U_5O_{13}, U_8O_{21}, and $U_{13}O_{34}$ intermediate between UO_2 & UO_3.

A turtle has 13 horn plates on its shell, 5 centered, 8 on the edges, 5 paw pins, and 34 backbone segments. There are 144 vertebrae in a Gabon snake, a hyena has 34 teeth, and a dolphin 233. Many spiders have 5 pairs of extremities, 5 parts to each extremity, and a belly divided into 8 segments carried by its 8 legs.

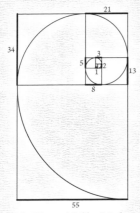

the Fibonacci Golden Spiral

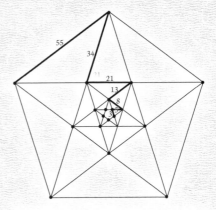

Fibonacci numbers approximate pentagram lengths

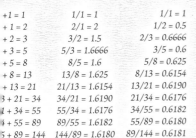

+1 = 1	1/1 = 1	1/1 = 1
+1 = 2	2/1 = 2	1/2 = 0.5
+2 = 3	3/2 = 1.5	2/3 = 0.6666
+3 = 5	5/3 = 1.6666	3/5 = 0.6
+5 = 8	8/5 = 1.6	5/8 = 0.625
+8 = 13	13/8 = 1.625	8/13 = 0.6154
+13 = 21	21/13 = 1.6154	13/21 = 0.6190
3 + 21 = 34	34/21 = 1.6190	21/34 = 0.6176
1 + 34 = 55	55/34 = 1.6176	34/55 = 0.6182
4 + 55 = 89	89/55 = 1.6182	55/89 = 0.6180
5 + 89 = 144	144/89 = 1.6180	89/144 = 0.6181

each term is the sum of the previous two

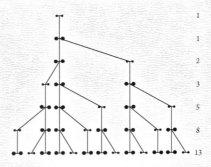

numbers of breeding pairs of rabbits

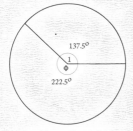

the golden angle, 360°/Φ²

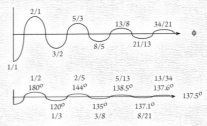

Fibonacci ratios converge on the golden ratio

PHYLLOTAXIS PATTERNS
leaves on a stem

Emerging as a science in the 19th century, phyllotaxis has been extended to the spiral patterns of seeds in a sunflower head, petals in the daisy, scales of pine cones, cacti areoles, and other patterns exhibited in plants. In the 15th century DaVinci [1452-1519] observed that the spacing of leaves was often spiral in arrangement. Kepler [1571-1630] later noted the majority of wild flowers are pentagonal, and that Fibonacci numbers occur in leaf arrangement.

Appropriately, in 1754 Charles Bonnet coined the name phyllotaxis from the Greek *phullon* "leaf" and *taxis* "arrangement." Schimper [1830] developed the concept of the divergence angle of what he called the "genetic" spiral, noticing the presence of simple Fibonacci numbers. The Bravais brothers [1837] discovered the crystal lattice and the ideal divergence angle of phyllotaxis: $137.5^\circ = 360^\circ/\Phi^2$.

The diagram by Church (*top row opposite*) shows the main features of spiral phyllotaxis. As the seed head expands, new primordia are formed at angles of 137.5°. In the seventh item we can see the Archimedean spiral which connects the growth. The diagrams below (*after Stewart*) show primordia plotted at angles of 137.3°, 137.5° and 137.6°. Only the precise angle produces a perfect packing.

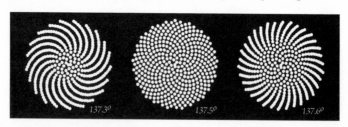

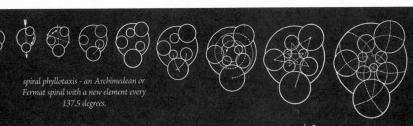

spiral phyllotaxis - an Archimedean or
Fermat spiral with a new element every
137.5 degrees.

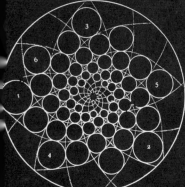

13:8 phyllotaxis - thirteen spirals one way,
eight the other way, known as parastichies.

34:21 phyllotaxis, with dots indicating the smaller-
numbered more highly curved spirals.

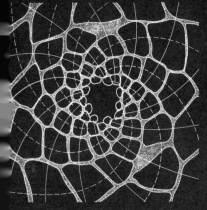

13:8 spiral phyllotaxis in a section of monkey puzzle tree

a seed head displaying 34:21 spiral phyllotaxis

ORDER BEHIND DIVERSITY
she loves me, she loves me not

Despite its seemingly endless variety and diversity, nature employs only three basic ways to arrange leaves along a stem: *disticious*, like corn, *decussate* or *whorled*, such as mint, and the most common, *spiral* phyllotaxis (for about 80% of the 250,000 different species of higher plants), where the divergence (rotation) angle between leaves has only a few values, these being close Fibonacci approximations to the golden angle, 137.5°. This pattern aids photosynthesis, each leaf receiving maximum sunlight and rain, efficiently spirals moisture to roots, and gives best exposure for insect pollination.

Opposing spirals of seeds in a sunflower generally appear as adjacent Fibonacci numbers, typically, 55:34 (1.6176) or 89:55 (1.6181). Scales of pinecones are typically 5:3 (1.6666) or 8:5 (1.6). Artichokes likewise display 8 spirals one way, 5 the other. Pineapples have three spirals, often 8, 13 and 21 each (*below*), where 21:13:8 approximates $\Phi : 1 : 1/\Phi$, with 21:13 (1.6153), 13:8 (1.625), and 21:8 (2.625), aspiring toward Φ^2, or $\Phi + 1$. Similarly, the pussy willow branch spirals 5 times with 13 buds appearing.

Next time you are out in the park, woods, countryside, or along a trail, take a moment to examine the petals on a daisy, count the spirals on a pinecone, or note the buds on a pussy willow.

8 gradual   *5 gradual*   *13 medium*   *8 steep*   *21 steep*

2/1   2/1   3/1   3/2   5/2   5/3   8/3   8/5

180°   180°   120°   240°   144°   216°   135°   225°

Simple phyllotaxis deriving from Fibonacci numbers - in each case a:b, a leaves are produced in b turns, meaning that the leaf divergence angle is (b/a)360°. As a and b increase, the divergence angles approach 137.5° and 222.5°.

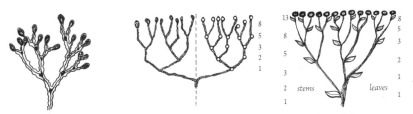

A frond of brown alga (left), with a schematic (center) showing the number of bifurcations exhibiting Fibonacci numbers. Sneezewort (right) also displays Fibonacci numbers in the count of its stems and leaves as it grows.

The three phyllotactic patterns, disticious, whorled, and spiral (after Ball), flanked by two Fibonacci spiral examples. The plant on the left displays 8 leaves in 5 turns, the pussy willow on the right produces 13 buds in exactly 5 turns.

15

LUCAS NUMBER MAGIC
integers perfectly formed from irrationals

In addition to the Fibonacci numbers, nature occasionally uses another series, named after Edouard Lucas. Lucas numbers (2, 1, 3, 4, 7, 11, 18, 29, 47, 76, 123, 199....) are similar to Fibonacci numbers in that they are additive (each new number is the sum of the previous two numbers), and multiplicative (each new number approximates the previous number multiplied by the modular Φ). In fact any additive series will converge on the golden ratio, the Fibonacci and Lucas series just do it the quickest. Note that the first four integers (*the basis of the Tetraktys - see page 55*) are all Lucas numbers.

What is fascinating about the Lucas numbers is that they are formed by alternately adding and subtracting the golden powers of Φ and its reciprocal $1/\Phi$, the two irrational parts either zipping together or peeling apart to form the integers (*opposite top*). These are not approximations, but absolutely exact! This extraordinary feature may be extended to the construction of Fibonacci numbers (*lower opposite*). Incredibly, it turns out that all integers can be constructed out of golden section powers, providing us with a tantalizing new way of constructing mathematics: integers are secretly hiding their component golden powers.

Together with Fibonaccis, Lucas numbers (though more rare) are sometimes found in the phyllotactic patterns of sunflowers (at times as much as 1 in 10 in some species), and in certain cedars, sequoias, balsam trees, and other species.

In general, the Lucas divergence angle of $99.5° = 360°/(1 + \Phi^2)$ occurs in 1.5% of observed phyllotactic plant patterns, as compared to 92% for the Fibonacci driven divergence angle (*see title page*).

16

0	$2 = \Phi + 1/\Phi^2$	=	$1.61803398... + 0.38196601...$	
1	$1 = \Phi - 1/\Phi$	=	$1.61803398... - 0.61803398...$	
2	$3 = \Phi^2 + 1/\Phi^2$	=	$2.61803398... + 0.38196601...$	
3	$4 = \Phi^3 - 1/\Phi^3$	=	$4.23606797... - 0.23606797...$	
4	$7 = \Phi^4 + 1/\Phi^4$	=	$6.85410196... + 0.14589803...$	
5	$11 = \Phi^5 - 1/\Phi^5$	=	$11.09016994... - 0.09016994...$	
6	$18 = \Phi^6 + 1/\Phi^6$	=	$17.94427191... + 0.05572808...$	
7	$29 = \Phi^7 - 1/\Phi^7$	=	$29.03444185... - 0.03444185...$	
8	$47 = \Phi^8 + 1/\Phi^8$	=	$46.97871376... + 0.02128623...$	
9	$76 = \Phi^9 - 1/\Phi^9$	=	$76.01315561... - 0.01315561...$	
10	$123 = \Phi^{10} + 1/\Phi^{10}$	=	$122.9918693... + 0.0081306...$	
11	$199 = \Phi^{11} - 1/\Phi^{11}$	=	$199.00502499... - 0.00502499...$	

$$7$$
$$=$$
$$G^4 + L^4$$

G^4	L^4
6 .	0 .
8	1
5	4
4	5
1	8
0	9
1	8
9	0

The Lucas Series: Even-termed members are formed by the addition of the greater and lesser powers of the golden section, odd-termed members by subtraction. Notice how the decimals in the odd terms are perfectly sliced off.

The number 7 is formed by zipping together the fourth powers of Φ and $1/\Phi$. Notice how the decimals sum to 9.

Fib. no.

2	$1 = \dfrac{\Phi^2 + 0}{\Phi^2}$	$= \Phi^0 + 0/\Phi^2$	$= G^0$	$= 1$	
3	$2 = \dfrac{\Phi^3 + 1}{\Phi^2}$	$= \Phi^1 + 1/\Phi^2$	$= G^1 + L^2$	$= 1.61803398... + 0.38196601...$	
4	$3 = \dfrac{\Phi^4 + 1}{\Phi^2}$	$= \Phi^2 + 1/\Phi^2$	$= G^2 + L^2$	$= 2.61803398... + 0.38196601...$	
5	$5 = \dfrac{\Phi^5 + 2}{\Phi^2}$	$= \Phi^3 + 2/\Phi^2$	$= G^3 + 2L^2$	$= 4.23606797... + 0.76393202...$	
6	$8 = \dfrac{\Phi^6 + 3}{\Phi^2}$	$= \Phi^4 + 3/\Phi^2$	$= G^4 + 3L^2$	$= 6.85410196... + 1.14589803...$	
7	$13 = \dfrac{\Phi^7 + 5}{\Phi^2}$	$= \Phi^5 + 5/\Phi^2$	$= G^5 + 5L^2$	$= 11.09016994... + 1.90983005...$	
8	$21 = \dfrac{\Phi^8 + 8}{\Phi^2}$	$= \Phi^6 + 8/\Phi^2$	$= G^6 + 8L^2$	$= 17.94427191... + 3.05572808...$	

Like the Lucas series, Fibonacci numbers can be expressed in terms of powers of the golden section. Notice the Fibonacci numbers reappearing in the equations - these can be further collapsed into golden power terms by repeatedly using the same technique.

ALL CREATURES
the divine symphony of life

Nature exhibits an array of beautiful and wondrous forms. Plants, trees, insects, fish, dogs, cats, horses and peacocks all display a poetic interplay between symmetry and asymmetry. Golden relations are often displayed through golden rectangles (*see beetles and fish opposite, after Doczi*), and their subsequent sectioning into component squares and smaller golden rectangles. This perpetuates the ratios of the original whole into its self-similar parts, reflecting the Φ:1:1/Φ proportional symmetry that we call the divine proportion. As Schwaller de Lubicz stated in *The Temple of Man*, "The impulse of all movement and all form is given by Φ."

The prevalence of natural pentagonal forms may result from the symphony of golden relationships in the pentagon and pentagram (*below and center row opposite, from Colman*). Many marine animals, like starfish, exhibit 5-fold form. Sometimes, as in a passion flower, the form is decagonal, one pentagon superimposed upon another.

Even the building blocks for life, ammonia (NH_3), methane (CH_4), and water (H_2O) all have internal bond angles which approximate the internal 108° angle of a pentagon.

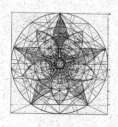 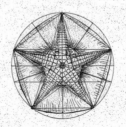 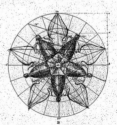

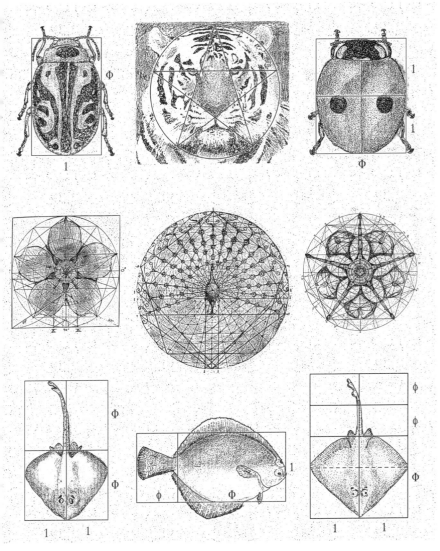

PHI IN THE HUMAN BODY
in the image of the divine

For over 20 years my students have been subjected to in-class measurements where their height relative to the location of their navel was determined. The purpose was to see whether the navel actually divides the body in the golden section, as some suggest the ancient *Canon of Polyclitus* purported. Over the years there have been a few who have exhibited what appears to be a perfect golden cut, however, in performing the calculations, the majority fit very closely into whole number Fibonacci approximations, particularly, the 5:3 range, and occasionally 8:5.

The golden ratio manifests throughout the human body. The three bones of each of your fingers are in golden relationship, and the wrist divides the hand and forearm at the golden section. Fibonacci numbers appear in your teeth, which sum to 13 in each quarter of your mouth over a lifetime, divided into a childhood 5 and an adult 8. The journey from child to adult also contains another surprise: a baby's navel (representing its past) is at its midpoint, and its genitals occur at the golden point, but when fully grown these reverse, as an adult's midpoint is at the genitals (the future) with the navel approximating the golden section (*opposite lower left*).

In Da Vinci's drawing of a head (*opposite top right*) a golden rectangle frames the face, and positions the eyes, nose and mouth.

Below we see Durer's drawings of less than golden faces.

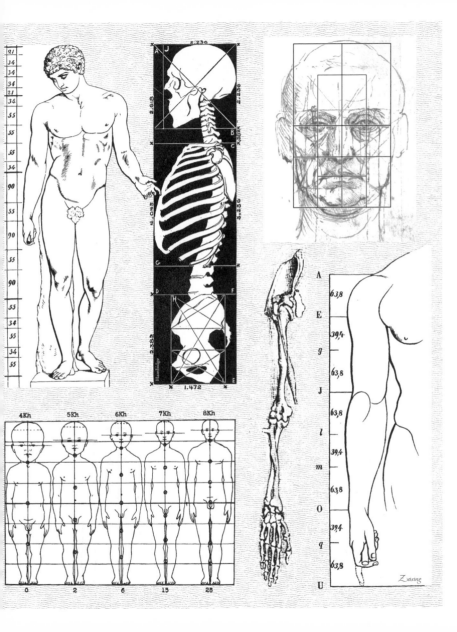

GROWTH & DIMINUTION
through the looking glass

Nature pulses with cycles and rhythms of increase and decrease. Heraclitus, a Presocratic influence on Plato, noted: "The way up and the way down are one and the same." Observe the waxing and waning of the moon, the circle of the year, the interplay of day and night, the breath of the tides, the systole and diastole of the beat of the heart, and the expansion and contraction of the lungs. The explosive growth of a star is often followed by implosion, and the negative entropy in the ordered organization of life is balanced by the positive entropy of disorder and death.

In chaos theory, the golden section governs the chaos border, where order passes into and emerges out of disorder (*see appendix V*). Demanding simplicity and economy, nature appears to require an accretion and diminution process that is simultaneously additive and multiplicative, subtractive and divisional. This demand is satisfied perfectly only by the golden section powers, and in practice by Fibonacci and Lucas approximations.

In the table (*opposite top*), notice how we can move upward in growth by both addition and multiplication, and move down, diminishing, by subtraction and division. The fulcrum is Unity acting as the geometric mean in golden relationship to both the increase of the deficient Lesser and decrease of excessive Greater.

Think of an oak tree. It shoots up as fast as it can from an acorn, only to slow, mature and fractalize its space toward a limit, becoming a new relative unity, what Aristotle called an entelechy, the form it grows into. Like Alice in Wonderland, nature simultaneously grows and diminishes to relative limits.

n	Greater	Mean	Lesser
7	Φ^7	Φ^6	Φ^5
6	Φ^6	Φ^5	Φ^4
5	Φ^5	Φ^4	Φ^3
4	Φ^4	Φ^3	Φ^2
3	Φ^3	Φ^2	Φ
2	Φ^2	Φ	1
1	Φ	1	$1/\Phi$
0	1	$1/\Phi$	$1/\Phi^2$
-1	$1/\Phi$	$1/\Phi^2$	$1/\Phi^3$
-2	$1/\Phi^2$	$1/\Phi^3$	$1/\Phi^4$
-3	$1/\Phi^3$	$1/\Phi^4$	$1/\Phi^5$
-4	$1/\Phi^4$	$1/\Phi^5$	$1/\Phi^6$
-5	$1/\Phi^5$	$1/\Phi^6$	$1/\Phi^7$
-6	$1/\Phi^6$	$1/\Phi^7$	$1/\Phi^8$
-7	$1/\Phi^7$	$1/\Phi^8$	$1/\Phi^9$

growth - the way up ->

diminution - the way down ->

The Golden Series shown opposite displays the unique simultaneous additive and multipicative qualities of the Golden Section.

Multiplication:
$$G_{n+1} = G_n \times \Phi$$
Addition:
$$G_{n+1} = G_n + M_n = G_n + G_{n-1}$$

Division:
$$G_{n-1} = G_n / \Phi$$
Subtraction:
$$G_{n-1} = M_n = G_n - L_n = G_n - G_{n-2}$$

These equations may be extended for Lesser and Mean values.

Each term is simultaneously the sum of the preceding two and the product of the previous term multiplied by Φ.

So $\Phi^4 = \Phi^2 + \Phi^3 = \Phi^2 \times \Phi^2 = \Phi^3 \times \Phi$

No other number behaves likes this, fusing addition and multiplication.

G	M	L		G	M	L
144	89	55		322	199	123
89	55	34		199	123	76
55	34	21		123	76	47
34	21	13		76	47	29
21	13	8		47	29	18
13	8	5		29	18	11
8	5	3		18	11	7
5	3	2		11	7	4
3	2	1		7	4	3
2	1	1		4	3	1
1	1	0		3	1	2
Fibonacci				*Lucas*		

The Fibonacci approximate geometric mean is corrected alternately by +1 or -1 under the square root. So 3 is the approximate geometric mean of 2 and 5, as $\sqrt{[(2 \times 5)-1]} = \sqrt{9}$, and 5 is the approximate geometric mean of 3 and 8, $= \sqrt{[(3 \times 8)+1]} = \sqrt{25}$.

The Lucas approximate geometric mean is corrected alternately by +5 or -5 under the square root. So 4 is the approximate geometric mean of 3 and 7, $= \sqrt{[(3 \times 7)-5]} = \sqrt{16}$, and 7 is the approximate geometric mean of 4 and 11, $= \sqrt{[(4 \times 11)+5]} = \sqrt{49}$.

EXPONENTIALS AND SPIRALS
an extended family of wonderful curves

In nature gnomonic growth occurs through simple accretion. It produces the beautiful logarithmic spiral growth we see in mollusks, which constantly add new material at the open end of their shells. Importantly, the shell grows in size, increasing in length and width, without varying its proportions. This accretive process, also used by crystals, is the simplest law of growth.

The golden spiral, derived from Fibonacci numbers (*as shown on the cover*), and from the arm of a pentagram (*below*), is a member of the family of logarithmic spirals. These also go by the name of growth spirals, equiangular spirals, and sometimes *spira mirabilis*, "wonderful spiral." When a spiral is logarithmic the curve appears the same at every scale, and any line drawn from the center meets any part of the spiral at exactly the same angle for that spiral. Zoom in on a logarithmic spiral and you will discover another spiral waiting for you. They are to be contrasted with Archimedean spirals, which have equal-spaced coils, like a coiled snake or hose.

Nature uses numerous different logarithmic spirals in leaf and shell shapes, cacti and seed-head phyllotaxis, whirlpools and galaxies. Many can be approximated using a family of golden spirals derived from equal divisions of a circle (*see opposite after Coates & Colman*).

  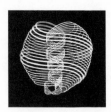

GOLDEN SYMMETRY
proportion from asymmetry

Nature presents us with a wonderful holographic portrait, where the smaller portions mirror the whole (cosmos) itself. Recognizing that structural self-similarity connects, or binds, what he called the hidden "implicate order" to the outer "explicate order," physicist David Bohm remarked: "The essential feature of quantum interconnectedness is that the whole universe is enfolded in everything, and that each thing is enfolded in the whole."

As we have seen, this marriage of the whole and its parts is elegantly accomplished via proportional symmetry, and in particular it is most efficiently produced by the golden section. This simple cut appears to be the driving impulse of nature itself, fractalizing with self-similarity into all the parts, and driving the growth process through spiraling golden angles and Fibonacci numbers.

It is the asymmetric push, the dynamic energy of the golden ratio manifesting as life, form and consciousness that provides the impetus to the rhythmic swing, the initial push of the pendulum.

The theme is explored in John Michell's painting, *The Pattern* (*opposite*). Concerning Intelligible symmetry, Michell writes: "Socrates called it the 'heavenly pattern,' which anyone can discover, and once they have found it they can establish it in themselves."

Someone must have hit the right note, because everything suddenly began falling into place.

PHI IN HUMAN CULTURE
sympathetic magic - as above, so below

A careful comparative study of cultures, their art, architecture, religion, mythology and philosophy, often reveals that, like phyllotaxis, the multiplicity and diversity of styles and types are underlain by strikingly simple universal principles. Plato maintained that the goal of aesthetics is not simply to copy nature, but rather to peer deeply into her, penetrating her tapestry to understand and employ the sacred ratios and proportions at work in her beautifully simple but divine order.

Concerning this, Plotinus [205-270] wrote:"The wise men of old, who made temples and statues in the wish that the gods should be present to them, looking to the nature of the All, had in mind that the nature of soul is easy to attract, but that if someone were to construct something sympathetic to it and able to receive a part of it, it would of all things receive soul most easily. That which is sympathetic to it is what imitates it in some way like a mirror able to catch the reflection of the form."

The designers of Beijing's early 15th century Forbidden City (*opposite*) used three equal and adjacent golden rectangles to frame their project, two of which enclose the moat. See if you can locate them. They then used the principle of *rabatment* to site and proportion further elements. In rabatment, squares are sectioned off inside golden rectangles to produce smaller golden rectangles, producing further guide lines (*see also the Tablet of Shamash, page 3*).

In the twelve pages which follow we examine in further detail some of the ways in which humanity has attempted to manifest, or craft, the divine forms of nature into the human environment.

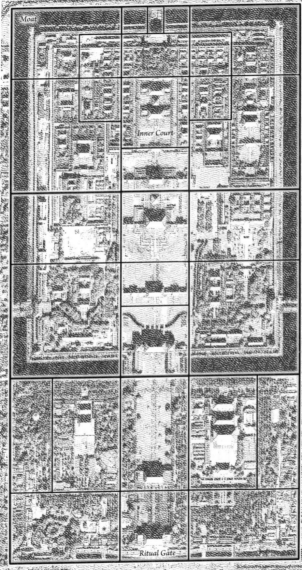

Moat

Inner Court

Ritual Gate

Tiananmen Square

ANCIENT OF DAYS
tombs, temples and pyramids

Like many ancient high cultures, the ancient Egyptians employed a sophisticated canon of number, measure and harmonic proportions in their magnificent monumental pyramids, temples and artwork. The simple ratios and grids they employed included the $\sqrt2$ diagonal of a square, the $\sqrt3$ bisection of an equilateral triangle, and the $\sqrt5$-based golden section, which appears both as golden ratio Fibonacci rectangles and in its pure pentagonal form. Fibonacci golden section approximations are suggested in the analyses of Hakoris chapel (8:5) and the Dendera Zodiac (5:3) (*opposite*). Moses built the Ark of the Covenant to a plan of 5:3 (2.5 x 1.5 cubits). Also note the pentagonal analyses apparent in the plan of the Osirian and the stunning statue of Menkaure (pharoanic builder of the smaller of the three Giza pyramids). The famous mask of Tutankhamen lends itself to a similar analysis.

Golden proportions and their Fibonacci approximations are found in Olmec sculpture (De La Fuente) and Mayan temple and art panel ruins in Palenque (C Powell). A 5:3 mandala (*below center*) regularly appears in Mesoamerican sculpture, architecture and codices (*below: Izapa Stela 89, & Olmec Monument 52, from Norman*).

Next time you are in the museum see what you can discover.

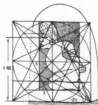
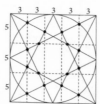
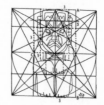

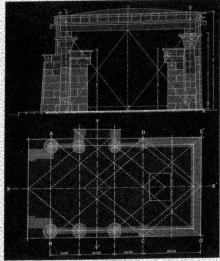
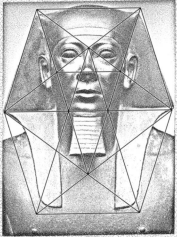

8:5 triangles in the chapel of Hakoris at Karnak (after Lauffray)

Lawlor's pentagonal analysis of the Osirian

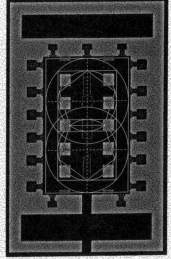
a 5:3 mandala underlying the Dendera Zodiac (after Harding)

pentagonal geometry in a bust of Menkaure

My Cup Runneth Over
half full or half empty

After careful study of Egyptian and Greek art, and what he called the architecture of plants, shells, man, and the five regular solids, Jay Hambidge developed a theory of dynamic symmetry, in which the same principle of self-similar growth of areas was found displayed throughout nature's living "form rhythms." He maintained that the dynamism was to be discovered in incommensurable lines that were commensurable in square, i.e., in area. Thus the ratios of √2, √3 and √5 became central to his work, with a special place for the whirling squares established in spiral rotation in the continued reduction of the golden rectangle.

Hambidge's geometrical analyses of various items of Greek pottery are shown below and opposite and his full set of designer rectangles is shown later (*appendix IV, page 56*). It has been claimed that Hambidge had a rectangle for everything, and that potters would have struggled to meet his exacting standards, but this is not to detract from his sincerity or his scholarly credentials.

There are certain lessons that may be drawn here – students of this subject may experience excessive enthusiasm over all things golden or, conversely, suffer complete skepticism and ossification.

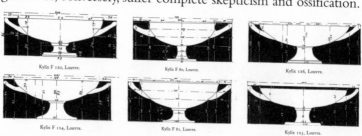

Kylix F 120, Louvre.

Kylix F 80, Louvre.

Kylix F 126, Louvre.

Kylix F 124, Louvre.

Kylix F 81, Louvre.

Kylix F 125, Louvre.

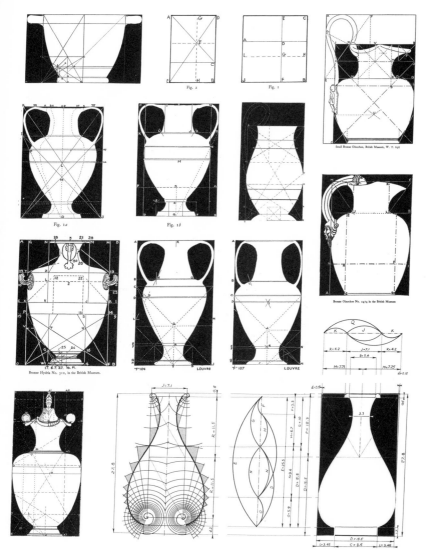

Fig. 2

Fig. 1

Small Bronze Oinochoe, British Museum, W. T. 656

Fig. 1*a*

Fig. 1*b*

17. 6. 7. 27. 16. A.
Bronze Hydria No. 312, in the British Museum.

"F" 104   LOUVRE

"F" 107   LOUVRE

Bronze Oinochoe No. 1474 in the British Museum

33

A SACRED TRADITION
old wine in new bottles

The philosophical and sacred number traditions of the Greeks and Romans were carefully carried into the new Christian religion as Jesus replaced Apollo and Hermes as the divine intermediary. The early tradition of the church placed emphasis upon the presence of Christ within, and the discovery of the Kingdom of Heaven in its divine proportion here on earth, in nature itself. Clement of Alexandria recognized Christianity as the "New Song," the sacred wine of the Logos occupying a new vessel.

Concerning the Logos (ratio or word), at the start of St. John's Gospel, 1:1, we read "In the beginning was the Logos, and the Logos was with God and the Logos was God." The only ratio that is simultaneously one and with one is the golden section.

Scripture with its symbolical and allegorical meaning can only be fully understood through a study of sacred number. By the science of gematria, the name Jesus ΙΗΣΟΥΣ sums to 888, Christ ΧΡΙΣΤΟΣ is 1480, and the two together 2368. These three names are in the golden proportion 3:5:8, with Christ the golden mean.

In Roman and Christian architecture the golden section was again used alongside integer ratios, and geometric diagonals √2, √3 and √5. Some examples are shown (*opposite and below*).

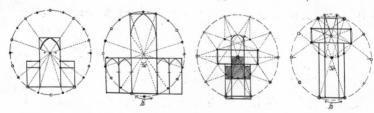

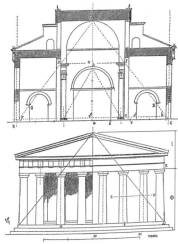

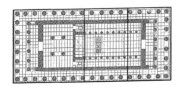

Clockwise from above: i) Portal relief over the south door of Chartres Cathedral displaying hidden pentagonal geometry (after Schneider). ii) Equilateral and 'Egyptian' 8:5 triangles in the Basilica of Constantine (Viollet-le-Duc). iii) 8:5 triangle defining the Parthenon (Viollet-le-Duc). iv) The plan of the Parthenon is a √5 rectangle, i.e., a square and two golden rectangles. v) Corinthian column capital displaying hidden pentagonal symmetry (after Palladio). vi) The Duomo, Florence, designed by Brunelleschi, showing golden rectangle relationships.

Opposite page: Moessel's decagonal analyses of the plans of Gothic churches and cathedrals embodying numerous golden proportions.

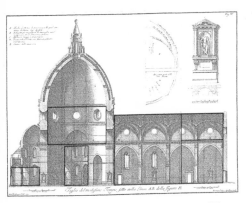

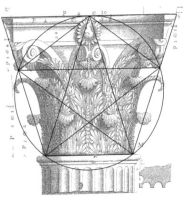

35

PHI IN PAINTING
further Da Vinci secrets

By carefully linking the ratios and proportions of a work of art, ensuring that the parts reflect and synchronize with the whole, a painter can create an aesthetic, dynamic, living embodiment of the harmonic and symmetrical principles lying behind nature itself.

Like the groundplan of the Parthenon (*previous page*), Leonardo DaVinci's painting of *The Annunciation* (*below*) employs a √5 framing rectangle (*below*). Using rabatment this is divided into one large square, and two golden rectangles which have both been further divided into a small square and a small golden rectangle. The device defines the main areas of the painting. In fact, in all the examples shown here the horizon is at the golden section of their height.

It is also not uncommon for artists to frame their pictures in 3:2, or 5:3 rectangles, simple Fibonacci approximations. Salvador Dalí's *Sacrament of the Last Supper* is a good example of the use of 5:3.

We can clearly see the aesthetic quality rendered by the combined asymmetry and proportional symmetry of the golden section.

Clockwise from top left: i) *The Virgin of the Rocks, Leonardo Da Vinci.* ii) *Alexander being lowered in a barrel, from The Alexander Romance (after Schneider).* iii) *The Beach, Vincent Van Gogh.* iv) *The Birth of Venus, Sandro Botticelli* v) *The Baptism of Jesus, Jean Colombe.*

37

MELODY & HARMONY
in search of the lost chord

Harmonics (number in time) was one of four disciplines studied in the Pythagorean Quadrivium, together with Arithmetic (pure number), Geometry (number in space), and Spherics (number in space and time). The golden section is a theme common to all.

In the Platonic tradition, the intention was to lift the soul out of the realm of mere opinion (*doxa*), by attunement with the ratios and proportions contained in the harmonies and rhythms of music. This allows the soul to pass into the Intelligible realm of knowledge (*episteme*), moving through the realm of mathematical reasoning (*dianoia*) up into direct intuition (*noesis*) of the world of pure Forms, the ratios themselves.

The structure of both rhythm and harmony is based upon ratio. The most simple and pleasing musical intervals, the octave (2:1) and the fifth (3:2), are the first Fibonacci approximations to the golden section. The series continues with the major and minor sixths (5:3 and 8:5). The scale itself holds the next step (13:8), for astonishingly, if we include the octave, musicians play eight notes in a scale, taken from thirteen chromatic notes. Finally, simple major and minor chords consist of the 1st, 3rd, 5th and 8th notes of the scale.

The golden section has been used by composers from Dufay (*opposite top after Sandresky*) to Bach, Bartok, and Sibelius, as a way of structuring a work of music. Russian musicologist Sabaneev discovered in 1925 that the golden section particularly appears in compositions by Beethoven (97% of works), Haydn (97%), Arensky (95%), Chopin (92%, including almost all of his *Etudes*), Schubert (91%), Mozart (91%), and Scriabin (90%).

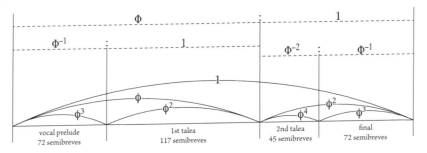

the vocal prelude of Dufay's Vasilissa, ergo gaude, composed around 1420

the structure of the whole Vasilissa is based on the golden section

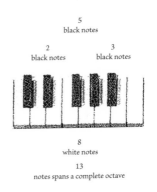

5
black notes

2
black notes

3
black notes

8
white notes

13
notes spans a complete octave

Fibonacci numbers appear in the modern scale and in pure harmonic intervals like the octave (2:1), the fifth (3:2), and the major and minor sixths (5:3) and (8:5).

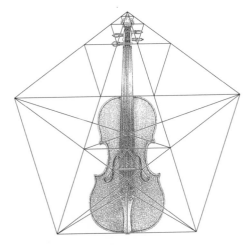

a Stradivarius violin

ALL THAT GLISTERS
is not gold

Today we consistently find ourselves reaching for that plastic card in our wallets and handbags. Most credit cards measure 86mm by 54mm, almost exactly an 8:5 rectangle and one of the most common Fibonacci approximations to the golden rectangle.

Because of its aesthetic qualities, embodied in its unique ability to relate the parts to whole, golden ratios are used in the design of many modern household items, from coffeepots, cassette tapes, playing cards, pens, radios, books, bicycles, and computer screens, to tables, chairs, windows and doorways. It even comes into literature, in the page layout of medieval manuscripts (*bottom right*) and as the small winged *Golden Snitch* in the Harry Potter stories.

Other important rectangles also find their way into our daily lives. The continuous geometric proportion most perfectly expressed in the golden series is mimicked in the International Standard Paper Size, which employs the continuous geometric proportion of $2:\sqrt{2}::\sqrt{2}:1$. Whereas removing a square from a golden rectangle produces another golden rectangle, folding a $\sqrt{2}$ rectangle in half produces two smaller $\sqrt{2}$ rectangles. Thus folding a sheet of A3 $(2:\sqrt{2})$ in half, gives you two sheets of A4 (each $\sqrt{2}:1$).

Golden dividers are a useful tool to have lying around the house (*beside the calculator opposite*). They can be made at any size, and opened to produce the golden section in any object that you are curious about. They are relatively easy to construct: simply mark three equal rods with the golden section, drilling holes in two of them at the mark and cutting the third. Observe the example shown, fasten in four places, and sharpen the tips to complete.

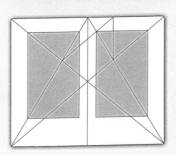

The Golden Chalice
a marriage of roots

Plato said learning is remembrance. The teacher acts as a midwife, and by close communion with the student, passes a spark (of resonance) lighting a flame, resulting in the birth of the innate idea. Contemplation of a drawing assists this process.

The construction below demonstrates how $\sqrt{3}$ is derivable as the hypotenuse of a right-angled triangle with Φ and $1/\Phi$ as legs. The Golden Chalice (*v., opposite*) combines this $\sqrt{3}$ revelation along with $\sqrt{2}$ derived from $\sqrt{\Phi}$ and $1/\Phi$ as legs. Critchlow's Kairos drawing (*vi., opposite*) derives a pentagram (with its Φ and $1/\Phi$) from a circle and a $\sqrt{3}$ equilateral triangle. All these results are exact!

Critchlow, concerned about the qualitative, ethical aspects of sacred geometry, writes: *"We are born into a world which appears as an indefinite dyad, a duality, a 'myself' and 'others,' until such time as we reach a maturity which can be called 'relationship.' This reveals itself as the unity that is the true case and we can realize it through the 'golden mean' of people's relationship with all others, including the environment. In the Kairos diagram, the golden mean is linking the trinitarian equilateral triangle to the life-emblem of the five-pointed star."*

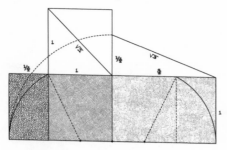

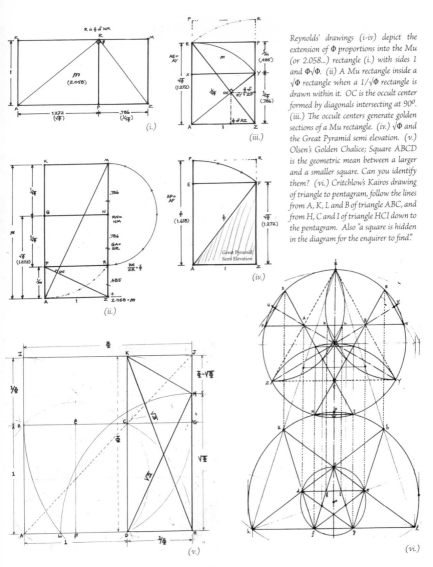

Reynolds' drawings (i-iv) depict the extension of Φ proportions into the Mu (or 2.058...) rectangle (i.) with sides 1 and Φ√Φ. (ii.) A Mu rectangle inside a √Φ rectangle when a 1/√Φ rectangle is drawn within it. OC is the occult center formed by diagonals intersecting at 90°. (iii.) The occult centers generate golden sections of a Mu rectangle. (iv.) √Φ and the Great Pyramid semi elevation. (v.) Olsen's Golden Chalice; Square ABCD is the geometric mean between a larger and a smaller square. Can you identify them? (vi.) Critchlow's Kairos drawing of triangle to pentagram, follow the lines from A, K, L and B of triangle ABC, and from H, C and I of triangle HCI down to the pentagram. Also "a square is hidden in the diagram for the enquirer to find".

GOLDEN POLYHEDRA
water, ether and the cosmos

The Golden Section plays a fundamental role in the structure of 3-D space, especially in the icosahedron and its dual, the dodecahedron (*opposite bottom right*), created from the centers of the icosahedron's faces. A rectangle drawn inside an icosahedron has edges in the ratio Φ:1 (or 1:ϕ) (*see below left and inside a cube opposite top*). The rectangles inside a dodecahedron are Φ^2:1 (or 1:ϕ^2) (*inside a cube lower opposite*). Nested inside an octahedron, the icosahedron cuts its edges in the ratio Φ:1 (*below center*). The exquisite early drawings by Kepler, DaVinci (*opposite*) and Jamnitzer [1508-1585] (*opposite page 1*) show their fascination with the Φ and root relationships in the 5 Platonic and 13 Archimedean polyhedra.

Continuing this theme is the truncated icosahedron (*opposite upper right*), known to us today as the structure of C_{60}, or the common soccer ball; the rectangle in this solid has edges in the ratio 3Φ:1. The icosidodecahedron (*opposite top left*) has a radius: edge of Φ:1 and the rhombic triacontrahedron (*opposite lower left*) is made of thirty Φ:1 diamonds.

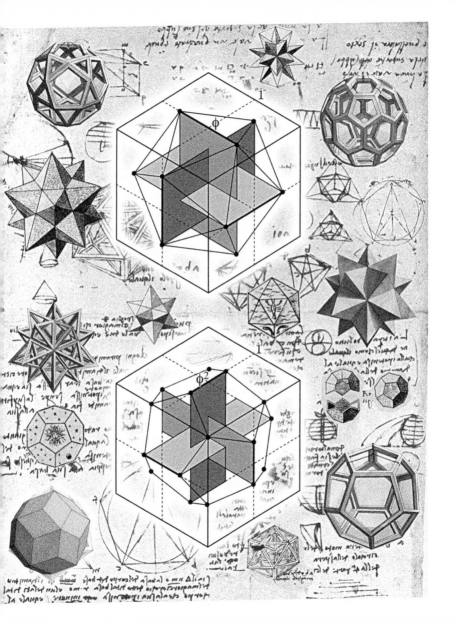

PHI IN THE SKY
Aphrodite's golden kiss

Not only the microcosm and mesocosm display a liking for the divine proportion. Golden ratios abound in the solar system, and, strangely, seem to occur particularly frequently around Earth. For example, both the relative physical sizes *and* the relative mean orbits of Earth and Mercury are given by Φ^2:1, or a pentagram to 99% accuracy (*opposite top left*).

Nothing, however, compares to the extraordinary relationship between Earth and our nearest planet, Venus which draws a beautiful five-fold rosette around us every eight years. Eight years on Earth is also thirteen Venusian years, the Fibonacci numbers 13:8:5 here appearing to connect space and time. Furthermore, Venus' perigee and apogee (her furthest and closest distances to Earth) are defined as Φ^4:1 to an accuracy of 99.99%, shown here as two nested pentagrams (*opposite, after Martineau*).

The two largest planets, Jupiter and Saturn, also produce a perfect golden ratio from Earth. Line them all up toward the Sun and a year later Earth is back where she started. Saturn will not have moved far and 12.85 days later Earth is again exactly between Saturn and the Sun. 20.79 days later Earth is found between the Sun and Jupiter. These synodic measures exist in space and time and relate as 1:Φ to 99.99% accuracy (*after Richard Heath*).

Moving yet further into the macrocosm, irrespective of whether or not they become reconstrued as dark energy stars, Paul Davies discovered that rotating black holes flip from a negative to a positive specific heat when the ratio of the square of the mass to the square of the spin parameter (rotation speed) equals Φ.

The two heavy circles show that the relative mean orbits of Earth and Mercury are close to $\Phi^2 : 1$. The sizes of the two planets are in the same ratio!

A technique for drawing the mean orbits of Earth and Venus. The two planets orbit the Sun at average distances in the ratio $(1 + 1/\Phi^2) : 1$.

The beautiful five-fold rosette pattern of comings and goings that Venus makes around Earth every eight years (or thirteen Venusian years).

Venus' furthest and closest distances from Earth, when she is in front of and behind the Sun, are precisely in the ratio $\Phi^4 : 1$.

RESONANCE & CONSCIOUSNESS
buddhas, shamans and microtubules

Consciousness is one of the great mysteries of humanity. Like life itself (*symbolized opposite center in the five-fold flower by Pablo Amaringo*), it may result from a resonance between the Divine (whole) and nature (the parts) exquisitely tuned by the amazing fractal properties of the golden ratio, allowing for more inclusive states of awareness.

Penrose (*the inventor of pentagonal tilings, opposite and below*) and Hameroff provocatively suggest that consciousness emerges through the quantum mechanics of microtubules. It is possible then that consciousness may reside in the geometry itself, in the golden ratios of DNA, microtubules, and clathrins (*opposite, by Gregory*). Microtubules, the structural and motile basis of cells, are composed of 13 tubulin, and exhibit 8:5 phyllotaxis. Clathrins, located at the tips of microtubules, are truncated icosahedra, abuzz with golden ratios. Perhaps they are the geometric jewels seen near the mouths of serpents by shamans in deep sacramental states of consciousness. Even DNA exhibits a Φ resonance. Each twist fits in a rectangle measuring in the Fibonacci ratio of 34:21 angstroms, and the cross-section through the molecule is decagonal.

Buddha said, "The body is an eye." In a state of Φ-induced quantum coherence, one may experience samadhi, cosmic conscious identification with the awareness of the Universe Itself.

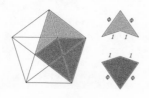

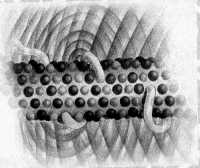

side view of a microtubule

looking into a microtubule

a double pentagon quintuplo flower

the ten-fold rosette cross-section of DNA

the soccer ball structure of a clathrin

THE PHILOSOPHER'S STONE
new vision and insight - a promise kept

We have come a long way, from a divided line to the essence of consciousness. The stated intention was to provide you with new vision and insight through examination of nature's greatest secret, the golden section, the most simple but profound asymmetric cut. Perpetuated throughout the cosmos at all levels, it marries endless variety with ordered proportional symmetry, unifying parts and whole from the large down to the little and back up again in a eurythmic symphony of form.

Together, on this journey, we just may have discovered the Pearl of Great Price, the precious Stone that transmutes base knowledge into golden wisdom. Next time you pick up a starfish, brush your teeth, admire a painting, see a pinecone, kick a soccer ball, gaze at the evening star, pick a flower, listen to some music, or even use your credit card, stop and think for a moment. You are a whole made up of lesser parts, and you are part of a greater whole.

This is nature's greatest secret. The golden section is interwoven into the very fabric of our existence, providing us with the means to resonate, to attune with successively broader stages of self-identity and unfoldment upon the path of return to the One.

It is humanity's duty to reconnect and resonate with this deep code of nature, beautifying our world and our relationships with eurythmic forms and golden standards of excellence. As nature does effortlessly, our duty is nothing less than to transmute our world, transforming it into the heavenly state of beauty and symbiotic peace that it was always intended to be.

APPENDIX I - PHI EQUATIONS

Expressions for phi:

The simultaneously additive and multiplicative nature of the golden section is expressed in the simple quadratic equation $a^2 - a = 1$ which has two solutions, one positive, one negative, Φ and $-\Phi^{-1}$:

$$a_1 = \frac{1 + \sqrt{5}}{2} \quad \text{and} \quad a_2 = \frac{1 - \sqrt{5}}{2}$$

thus $\Phi = \dfrac{\sqrt{5} + 1}{2} = 1.61803398874989484882..$

and $\dfrac{1}{\Phi} = \dfrac{\sqrt{5} - 1}{2} = 0.61803398874989484882..$

This also gives the following important identities:

$$\Phi = 1 + \frac{1}{\Phi} \quad \text{and} \quad \Phi = \sqrt{1 + \Phi}$$

Taking the first of these and repeatedly substituting for Φ produces the simplest continued fraction:

$$\Phi = 1 + \cfrac{1}{1 + \cfrac{1}{\Phi}} = 1 + \cfrac{1}{1 + \cfrac{1}{1 + \cfrac{1}{\Phi}}}$$

$$= 1 + \cfrac{1}{1 + \cfrac{1}{1 + \cfrac{1}{1 + \cfrac{1}{1 + \cfrac{1}{1 + \cdots}}}}}$$

while taking the second and repeatedly substituting for Φ produces the simplest nested radical:

$$\Phi = \sqrt{1 + \sqrt{1 + \Phi}} = \sqrt{1 + \sqrt{1 + \sqrt{1 + \Phi}}}$$

$$= \sqrt{1 + \sqrt{1 + \sqrt{1 + \sqrt{1 + \sqrt{1 + \cdots}}}}}$$

Approximate relationships between phi, pi and e:

The two formulae $\pi \cong 6/5\, \Phi^2$ and $\pi \cong 4/\sqrt{\Phi}$ may both be derived from the Great Pyramid. Note too the approximate formulae $e \cong \Phi^2 + \frac{1}{10}$ and the even more accurate $e \cong \frac{144}{55} + \frac{1}{10}$.

Trigonometrical functions involving phi:

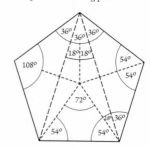

Angle θ	Sin θ	Cos θ	Tan θ
$18°$	$\dfrac{\sqrt{1 - 1/\Phi}}{2}$	$\dfrac{\sqrt{2 + \Phi}}{2}$	$\dfrac{\sqrt{1 - 1/\Phi}}{\sqrt{2 + \Phi}}$
$36°$	$\dfrac{\sqrt{2 - 1/\Phi}}{2}$	$\dfrac{\sqrt{1 + \Phi}}{2}$	$\dfrac{\sqrt{2 - 1/\Phi}}{\sqrt{1 + \Phi}}$
$54°$	$\dfrac{\sqrt{1 + \Phi}}{2}$	$\dfrac{\sqrt{2 - 1/\Phi}}{2}$	$\dfrac{\sqrt{1 + \Phi}}{\sqrt{2 - 1/\Phi}}$
$72°$	$\dfrac{\sqrt{2 + \Phi}}{2}$	$\dfrac{\sqrt{1 - 1/\Phi}}{2}$	$\dfrac{\sqrt{2 + \Phi}}{\sqrt{1 - 1/\Phi}}$

Linking Φ, e and i:

Richard Feynman noticed the equation, based on Euler's, that $e^{i\pi} = \Phi^{-1} - \Phi$. We also have the two results: $2\sin(i \ln \Phi) = i$, and $2\sin(\pi/2 - i \ln \Phi) = \sqrt{5}$.

The golden string:

Intimately connected to Φ is the infinite binary "rabbit" sequence, which never contains 00 or 111 and arises in many ways: 1011010110 1101011010 1101101011 0110101101 0110110101 1010110110 1011011010 1101011011 ... [see www.mcs.surrey.ac.uk]

APPENDIX II - FIBONACCI & LUCAS FORMULÆ

Definition of the Fibonacci Series:
$F_0 = 0, F_1 = 1$, then $F_{n+2} = F_{n+1} + F_n$

Early Fibonacci numbers:

n	0	1	2	3	4	5	6	7	8	9	10	11	12	13	14	15
F_n	0,	1,	1,	2,	3,	5,	8,	13,	21,	34,	55,	89,	144,	233,	377,	610,

16	17	18	19	20	21	22	23
987,	1597,	2584,	4181,	6765,	10946,	17711,	28657

24	25	26	27	28	29
46368,	75025,	121393,	196418,	317811,	514229 ...

Binet formula for Fibonacci numbers:
$F_n = (\Phi^n - (- \Phi^{-n})) / \sqrt{5}$

Cassini formula for Fibonacci numbers:
$(F_{n-1})(F_{n+1}) - (F_n)^2 = (-1)^n$

Negative termed Fibonacci numbers:
$F_{-n} = (-1)^{n+1} F_n$

Factors of Fibonacci numbers:

Every nth Fibonacci number is a multiple of F_n, so F_n is a factor of every nth Fibonacci number. Thus $F_3 = 2$ divides every 3rd Fibonacci number, meaning every third Fibonacci number is even; $F_4 = 3$ means every 4th Fibonacci number is divisible by 3, $F_5 = 5$ divides every 5th Fibonacci number, and $F_6 = 8$ divides every 6th Fibonacci number. Also if n is a factor of m, then F_n will be a factor of F_m.

Summing Fibonacci numbers:

$\sum F_n = F_{(n+2)} - 1$, ie the sum of the first n Fibonacci numbers is one less than the $n+2$nd Fibonacci number. Odd-numbered Fibonacci terms sum to the next even-numbered Fibonacci term while even-termed Fibonacci numbers sum to one less than the next odd-numbered term.

The squares of Fibonacci numbers:

$\sum (F_n)^2 = F_n F_{(n+1)}$ which means the sum of the squares of the first n Fibonacci numbers is equal to the product of the nth and the n+1th Fibonacci numbers. Also $(F_n)^2 = F_n (F_{(n+1)} - F_{(n-1)})$. The sum of the squares of two consecutive Fibonacci numbers $(F_n)^2 + (F_{n+1})^2 = F_{(2n+1)}$.

Definition of the Lucas Series:
$L_0 = 2, L_1 = 1$, then $L_{n+2} = L_{n+1} + L_n$

Early Lucas numbers:

n	0	1	2	3	4	5	6	7	8	9	10	11	12	13	14
L_n	2,	1,	3,	4,	7,	11,	18,	29,	47,	76,	123,	199,	322,	521,	843,

15	16	17	18	19	20	21	22
1364,	2207,	3571,	5778,	9349,	15127,	24476,	39603,

24	25	26	27	28	29
64079,	103682,	167761,	271443,	439204,	710647, ...

Binet formula for Lucas numbers:
$L_n = \Phi^n + (-\Phi^{-n})$

Cassini formula for Lucas numbers:
$(L_n)^2 - (L_{n+1})(L_{n-1}) = 5(-1)^n$

Negative termed Lucas numbers:
$L_{-n} = (-1)^n L_n$

The Cassini formulae:

As shown in the formula to the left, each Fibonacci number is the approximate geometric mean of its two adjacent numbers, alternately needing correction by +1 or -1, while above we see that each Lucas number is the approximated geometric mean of its two neighbors alternately corrected by -5 or +5. The two are further related by the expansion of Binet's formula as $\Phi^n = (L_n + F_n \sqrt{5})/2$.

Converting Fibonacci and Lucas numbers:

$L_n = F_{n+1} + F_{n-1}$, which is to say that the nth Lucas number is the sum of the n+1th and n-1th Fibonacci numbers. Related to this is the result $L_n = F_{n+2} - F_{n-2}$. We also have $L_n = F_n + 2F_{n-1}$, and the fact that any four consecutive Fibonacci numbers sum to a Lucas number. Finally there is the simple, elegant equation $F_{2n} = F_n L_n$, and the fact that $F_n + L_n = 2 F_{(n+1)}$.

Hyperbolic Fibonacci and Lucas functions:

From Binet's formula we can derive the fascinating pair of equations $L_{2n} = 2/\cosh(n \ln \Phi)$ and its successor $L_{2n+1} = 2 \sinh(n \ln \Phi)$. In 2003, Alexey Stakhov published the following two remarkable identities: $\sin F_n + \cos F_n = \sin F_{n+1}$, and $\sin L_n + \cos L_n = \cos L_{n+1}$.

APPENDIX III - THE INDEFINITE DYAD

Plato, as a Pythagorean, was under an inviolable oath not to reveal the deeper truths of the mystical Pythagorean mathematical order. Like Pythagoras, he spent considerable time in Egypt studying the mathematical mysteries with the priesthood there and in his writings intentionally conceals deeper truths under a shroud of mystery. As a teacher and writer, Plato practiced Socratic midwifery, presenting anomalous puzzles, problems, and incomplete solutions, along with significant hints, both within the Academy and in his dialogues. Readers needed to abduct (retroduct or hypothesize) a solution to the anomalous situation. Academy members were presented with problems such as the doubling of the cube, or giving an account of the heavens that would accommodate and explain the apparent irregular motions of the planets, thus "saving the appearances."

In the dialogues, Plato carefully selects several interrelated problems that are very subtly posed. Taken together they point to the great mystery of the Golden Section and its Reciprocal, none other than the Greater and Lesser of the Indefinite Dyad. It is clear from Aristotle and other members of the Academy that, in unwritten lectures, Plato more openly revealed the deep truths of how the One (or Good of the dialogues) combined with the Greater and Lesser of the Indefinite Dyad to produce the Hierarchy of Intelligible Forms, Mathematicals, and Sensible Particulars. In the *Metaphysics* [987b 19-22], Aristotle writes: "*Since the Forms are the causes of all other things… their elements were the elements of all things. As matter, the Great and Small were Principles; as substance, the One; for from the Great and Small, by participation in the One come the Forms, the Numbers.*" And yet even to the Academy members, the presentation was necessarily enigmatic. As Simplicius records in his *Commentary on Aristotle's Physics* [187a12], "*Plato maintained that the One and the Dyad were the First Principles, of Sensible Things as well. He placed the Indefinite Dyad also among the Objects of Thought and said it was Unlimited, and he made the Great and the Small First Principles and said they were Unlimited, in his lectures On the Good; Aristotle, Heraclides, Hestiaeus, and other associates of Plato attended these and wrote them down in the enigmatic style in which they were delivered.*"

In the *Parmenides* [133b], Plato presents the "*worst difficulty argument*": how can the immaterial Intelligible and material Sensible worlds have any contact or interaction? In the *Timaeus* [31b-32a], Plato makes it clear that continuous geometric proportion is the best of all bonds. This involves

there being an intermediate geometric mean relationship. He then gives us the so-called Lambda relationships of 1, 2, 4, 8 and 1, 3, 9, 27 [35b-36b]. In the *Republic* [509d], Plato asks us to divide the line unevenly, representing the Intelligible and Sensible Worlds. In effect he is telling us to generate a continuous geometric proportion between the whole and the parts with the simplest of all cuts, the Golden Cut. When we apply the same ratio, the Golden Cut, to the two segments, we now get the most interesting geometric proportion between the parts, $\Phi:1 :: 1:1/\Phi$. Unity becomes the geometric mean between Φ and $1/\Phi$. Therefore, the geometric relationship is $\Phi : 1 : 1/\Phi$. I argue that this is the Greater : Unity :: Unity : Lesser, or Greater : Unity : Lesser. Thus, the worst difficulty argument is answered through

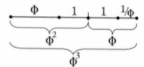

continuous geometric proportion. The Intelligible and Visible worlds are interlocked, interlaced, fused together through the magic of the Golden Mean in its Greater and Lesser relationships with Unity.

Furthermore, Plato in the *Timaeus* conspicuously omits the triangle required to construct the dodecahedron, even though he declares that this regular solid represents the Cosmos itself. This triangle would of course necessarily require him to overtly recognize the Golden Section. Yet he does provide the $\sqrt{2}$ triangle for the construction of the cube, and the $\sqrt{3}$ triangle for the construction of the tetrahedron, octahedron and icosahedron. He also notes that the $\sqrt{3}$ triangle is used to compose a 3rd triangle, the

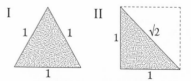

equilateral triangle. Plato's nephew Speusippus, who headed the Academy after Plato's death, wrote in *On Pythagorean*

Numbers (a fragment of which still survives) that the equilateral triangle with all sides equal represents Unity or One, the triangle with two sides equal represents Two, and the √3 triangle with three unequal sides represents Three.

Plato clearly calls the √2 and √3 triangles the most beautiful triangles. However, he then hints to the astute reader that: *"these then … we assume to be the original elements of fire and other bodies, but the principles which are prior to these God only knows, and he of men who is a friend of God."* Timaeus [53d-e] Thus, Plato suggests there may actually be principles prior to these triangles. In fact, I would suggest, the principles may be discoverable in the missing 4th triangle, the one necessary to construct the dodecahedron. Plato then goes on to say:

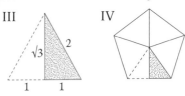

"…anyone who can point out a more beautiful form than ours for the construction of these bodies shall carry off the palm, not as an enemy, but as a friend. Now, the one which we maintain is the most beautiful of all the many triangles… is that of which the double forms a 3rd triangle which is equilateral. The reason of this would be too long to tell; he who disproves what we are saying, and shows that we are mistaken, may claim a friendly victory." Timaeus [54a-b]

It is particularly relevant that Plato names a 3rd triangle here, the equilateral triangle, that Speusippus has clearly identified as representing the One. The √2 triangle represents the Two, and the √3 triangle represents the Three. However, in Pythagorean language, one swears by the four-fold Tetraktys. And very suggestively, the *Timaeus* opens with the

cryptic statement: *"One, two, three, but where is the fourth, my dear Timaeus, of those who were yesterday my guests and are to be my entertainment today?"* One of the guests, the philosopher, has been taken ill. He is conspicuously absent, as Plato intends much more by these opening remarks. The fourth is now

the 4th triangle which is missing. It is the triangle that would build the dodecahedron, but it would require revealing the Golden Section. And not only that, it might reveal the very principles behind the Cosmos itself, namely, the One and the Indefinite Dyad of the Greater (Φ) and the Lesser (1/Φ) (*see The Golden Chalice, Fig. 5, p. 40-41, where √2 and √3 are derivable precisely from the Greater and the Lesser*).

The clincher comes in Alexander's *Commentary on the Metaphysics*, where he retains from Aristotle a very poignant observation regarding Plato: *"Thinking to prove that the Equal and Unequal [other names for the One and Indefinite Dyad] are first Principles of all things, both of things that exist in their own right and of opposites … he assigned equality to the monad, and inequality to excess and defect; for inequality involves two things, a great and a small, which are excessive and defective. This is why he called it an Indefinite Dyad - because neither the excessive nor the exceeded is, as such, definite. But when limited by the One the Indefinite Dyad, he says, becomes the Numerical Dyad."* And we know that when the One is added to the difference between the Greater and Lesser, it equals, not approximately, but exactly, Two. Thus, Greater − Lesser + Unity = Two; (Φ - 1/Φ) + 1 = 2.

Finally, it is also the way in which the Golden Section and its Reciprocal, together with Unity, set the standard of Proportion through the Geometric Mean relationship that is relevant. This pertains not only to Truth, the Reality of the Unfoldment of the Cosmos, but also to Beauty and Goodness. Thus in the *Statesman* [284a1-e8], Plato writes: *"It is in this way, when they preserve the standard of the Mean that all their works are Good and Beautiful…. The Greater and the Less are to be measured in relation, not only to one another [i.e., G:L = Φ²], but also to the establishment of the standard of the Mean [i.e., G:1 = Φ, and 1:L = Φ].… This other comprises that which measures them in relation to the moderate, the fitting, the opportune, the needful, and all the other standards that are situated in the Mean between the Extremes."* We now begin to see the extension to Aesthetics (Beauty) and Ethics (Goodness). Aristotle proposed, not surprisingly, the notion of the Golden Mean of Moderation between the extremes of too much and too little. For example, courage is preferable to the extremes of foolhardiness and cowardice.

Thus, the Indefinite Dyad in relation to Unity provides the basis for Truth, Beauty and Goodness. Johannes Kepler simultaneously concealed and revealed the nature of the great mystery, uttering the simple words: *"Geometry has two great treasures: one is the theorem of Pythagoras; the other, the division of a line into extreme and mean ratio. The first we may compare to a measure of gold; the second we may name a precious jewel."*

[this is an abbreviated version of a paper first appearing in *Nexus Journal of Architecture and Mathematics, vol. 4, no. 1. www.nexusjournal.com*]

APPENDIX IV - DESIGNER'S RECTANGLES

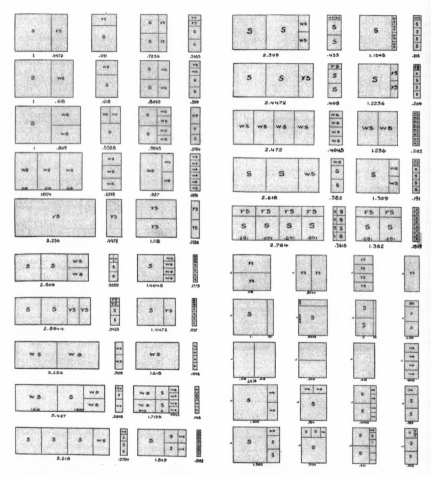

From Hambidge. Key: ws = whirling square (golden rectangle), s = square, v5 = root 5 rectangle

Appendix V - Golden Physics

David Bohm's deep platonic insights into nature's superimplicate, implicate & explicate orders may be combined with M.S. El Naschie's E-infinity (E^∞) theory. This models a harmonic production of quarks and elementary particles through a golden section centered Cantorian space-time. Bohm maintained that there is an inner, hidden implicate order (analogous to Plato's intelligible realm) lying behind the outer explicate order (Plato's sensible realm). He argued that this source of order and structure was discoverable in the so-called vacuum state, the zero-point energy field. In seminars at Birkbeck he asserted: "In one cubic centimeter of [so-called] empty space, the amount of energy is much greater than the total energy of all the matter in the known universe!" Matter is merely an "excitation on the virtual sea of the implicate order."

Reminiscent of Plato's prisoner watching the shadows on the cave wall, Bohm maintained that the fractal "discontinuity or sudden jumps at the quantum level may be considered as a shadow crossing the wall." M.S. El Naschie's contribution provides the detailed content for Bohm's platonic conceptual framework. Beginning with his 1994 paper, "Is Quantum Space a Random Cantor Set with a Golden Mean Dimension at the Core?" the E^∞ space-time theory provides a profound theoretical basis for the central role the golden section plays as the *"winding number"* in the harmonic manifestation of quark and subatomic particle masses through the continuous symmetry breaking of vacuum state fluctuations:

> *"The appearance of the Golden Mean, its inverse as well as its square value with both negative and positive signs as the frequency of vibration and mass-energy factor indicate that it is the simplest realistic unit from which a Hamiltonian dynamics can start developing a highly complex structure, a so-called nested vibration.... The Golden Mean plays a decisive role in nonlinear dynamical stability and chaotic systems as shown in the celebrated KAM theorem [Chaos Border of Kolmogorov, Arnold and Moser] and in high energy particle physics.... The KAM theorem asserts that the most stable periodic orbit is that which has an irrational ratio of resonance frequencies. Since the Golden Mean is ... the most irrational number ... the corresponding orbit is the most stable orbit.... In the view of string theory, particles are vibrating strings. Therefore to observe a particle, the corresponding vibration must be stable and that is only possible in the KAM interpretation which we call the VAK Cantorian theory of vacuum fluctuation, when the winding number corresponding to this dynamics is equal to the Golden Mean."* - El Naschie

El Naschie discovered that particle physics seen through the eyes of E^∞ appears to be "a cosmic symphony." The particles "are a rather non-complex function of the golden mean and its derivatives." The following E^∞ quark masses "are in excellent agreement with the majority of the scarce and difficult to obtain data about the mass of quarks. It takes only one look at these values for anyone to realize that they form a harmonic musical ladder."

CURRENT & CONSTITUENT QUARK MASS AS FUNCTIONS OF Φ AND $1/\Phi$.

Quark Flavor	Current Mass (MeV)	Constituent Mass (MeV)
Up	$2\,\Phi^2 = 5.236...$	$80\,\Phi^{-3} = 338.885...$
Down	$2\,\Phi^3 = 8.472...$	$80\,\Phi^{-3} = 338.885...$
Strange	$10\,\Phi^6 = 179.442...$	$10\,\Phi^8 = 469.787...$
Charm	$300\,\Phi^3 = 1,270.82...$	$20\,\Phi^9 = 1,520.263...$
Beauty (Bottom)	$10^3\,\Phi^3 = 4,236.067...$	$100\,\Phi^8 = 4,697.871...$
Truth (Top)	$10^4\,\Phi^3 = 42,360.679...$	$10^4\,\Phi^6 = 179,442.719...$

In the table below, notice the close agreement between the theoretical and the experimental values, and the interesting presence of the $5/2$ phyllotaxis and Lucas $7/4$ ratios. The E^∞ values of the fundamental constituents involved below are as follows: $\overline{\alpha}_0 = 20\Phi^4 = 137.0820...$ is the inverse Sommerfield electro-magnetic fine structure coupling constant. $\kappa = \Phi^{-3}(1 - \Phi^{-3}) = 0.12033988...$ is a Φ-based constant. $\overline{\alpha}_g = 10\Phi^3 = 42.3606797...$ is the theoretical value of the coupling constant $\overline{\alpha}_0$ at the point where three non-gravitational forces intersect. $\overline{\alpha}_{gs} = \overline{\alpha}_g/\Phi = 26.18033988... = (10\Phi^3)/\Phi = 10\Phi^2$. This is the inverse coupling constant at the super symmetric unification of all fundamental forces taking place at the Planck length of 10^{-33} cm.

SUBATOMIC PARTICLE MASS AS A FUNCTION OF Φ AND $1/\Phi$.

subatomic particle	theoretical mass (MeV)	experimental value (MeV)
e (electron)	$\sqrt{\overline{\alpha}_{gs}}/10 = \sqrt{(10\,\Phi^2)}/10$ $= 0.51166...$	0.511
n (neutron)	$20\,\Phi^8$ $= 939.574...$	939.563
P (proton)	$20\,\Phi^8 \cos(\pi/60)$ $= 938.28...$	938.27231
$\Pi^\pm$ (Π meson)	$\overline{\alpha}_0 + (5/2)$ $= 139.5820...$	139.57
Π^0	$\overline{\alpha}_0 - (5/2)$ $= 134.5820...$	134.98
Ω^-	$10[\,\overline{\alpha}_0 + (49/\Phi)]$ $= 1,673.657...$	1,672.43
Exi$^-$	$10[\,\overline{\alpha}_0 - (8/\Phi)]$ $= 1,321.377...$	1,321.32
Exi^0	$10[\,\overline{\alpha}_0 - (9/\Phi)]$ $= 1,315.197...$	1,314.9
μ (muon)	$\sqrt{(1000\,\Phi^5)} = 105.309$ or $(20+\kappa)(5+\Phi^{-3}) = 105.665...$	105.65839
η	$(4\,\overline{\alpha}_{gs})^2/20 = (40\,\Phi^2)^2/20$ $= 548.328... = m_\eta$	548.8
η'	$(7/4)\,m_\eta = (7/80)(4\,\overline{\alpha}_{gs})^2$ $= 959.5742755...$	957.5

Appendix VI - More Lucas Magic

In the early 1990's English researcher Robin Heath observed the strange fact that all of the marriage numbers of the Sun, Moon and Earth can be reduced to combinations of the operation of the Golden Section around the key numbers 18 and 19. The evidence for the favor that the Sun and Moon have for 18 and 19 is indicated both by the *Metonic Cycle*, which produces full moons at the same calendar dates 19 years later, and the *Saros Eclipse Cycle*, which results in similar eclipses repeating every 18 years.

In addition, the *Moon's Nodes*, the crossing points of the Solar and Lunar orbits as viewed from Earth, take 18.618 *years* to rotate once around the heavens, or 18 plus the Lesser golden mean. Heath made the startling observation that this same number squared produces 346.63, a highly accurate value for the *Eclipse Year* in *days*, which is the length of time it takes for the Sun to return to the same Moon's node (the nodes of course rotating slowly in the opposite direction to the Sun, Moon, and planets). Adding the magic number 18.618 to this produces 365.25, the number of days in a *Solar Year*, also 18.618 x 19.618. Finally, adding 18.618 again we obtain 383.87, which is *13 lunations*, or the number of days between 13 full moons, also 18.618 x 20.618.

Recalling that 18 is a Lucas number ($18 = \Phi^6 + \Phi^{-6}$), we may rephrase the already mentioned expressions thus:

Eclipse Year
$$= 18.618 \times 18.618 \text{ days}$$
$$= (\Phi^6 + \Phi^{-6} + \Phi^{-1})^2 \text{ days}$$

Solar Year
$$= 18.618 \times 19.618 \text{ days}$$
$$= (18 + \Phi^{-1})(18 + \Phi)$$
which reduced by Lucas magic
$$= (\Phi^6 + \Phi^{-6} + \Phi^{-1})(\Phi^6 + \Phi^{-6} + \Phi) \text{ days}$$

further expressed by Benjamin Bryton as
$$= \sum_{5,7,12} (\Phi^n + \Phi^{-n} + 1) = \sum_{6,8,13} (\Phi^n + \Phi^{-n})$$
(note here the structure of the scale)

13 Moons
$$= 18.618 \times 20.618 \text{ days}$$
$$= (\Phi^6 + \Phi^{-6} + \Phi^{-1})(\Phi^6 + \Phi^{-6} + \Phi^2) \text{ days}$$

Appendix VII - Phyllotaxis Angles

1/2	180°	elm, lime, birch, basswoods, cereals, grapes, some grasses.
1/3	120°	beech, hazel, alder, fiddleneck, blackberry, sedge, tulip, some grasses.
2/5	144°	oak, cherry, apple, holly, plum, apricot, coast live oak, California bay, pepper tree, manzanita, common groundsel, mustard, toyon, madrona.
3/8	135°	poplar, pear, weeping willow, rose, petty spurge, locust (thorn phyllotaxy), cabbage, radish, flax, plantains.
5/13	138.5°	almond, bottlebrush, pussy willow, spruce, jasmine, cranberry, leeks.
13/34	137.6°	pines, magnolia.

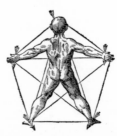

58